basic flavorings
Olive Oil

basic flavorings

Olive Oil

Clare Gordon-Smith

photography by

James Merrell

COURAGE
BOOKS

AN IMPRINT OF RUNNING PRESS
PHILADELPHIA • LONDON

Art Director **Jacqui Small**

Art Editor **Penny Stock**

Design Assistant **Mark Latter**

Editor **Elsa Petersen-Schepelern**

Photography **James Merrell**

Food Stylist **Clare Gordon-Smith**

Stylist **Sue Skeen**

Production Consultant **Vincent Smith**

**Our thanks to Christine Walsh and Ian Bartlett,
Jenny Merrell, Peter Bray, and Robert Roseman.**

© 1996 by Ryland Peters & Small

Produced by Mandarin Offset
Printed and bound in China

10 9 8 7 6 5 4 3 2 1
Digit on the right indicates the number of this printing

Library of Congress Cataloging-in-Publication number
96-67169

ISBN 1-56138-777-0

This edition published in the United States of America in 1996 by
Courage Books, an imprint of
Running Press Book Publishers
125 South Twenty-second Street
Philadelphia, Pennsylvania 19103-4399

Notes:
Ovens should be preheated to the
specified temperature. If using a
convection oven, adjust times and
temperature according to
manufacturer's instructions.

All pickles and preserves should be
processed in a boiling water-bath
canner according to USDA guidelines.

The FDA has raised concerns about
potential health hazards of homemade
flavored oils. The problem occurs when
bacteria that exists naturally in some
flavoring items are introduced to the
airless environment of oil. If you choose
to make flavored oil, the following steps
are recommended:
. Gently heat flavoring ingredients,
such as crushed garlic and herbs, in the
oil. Do not allow items to stand in
unheated oil.
. Make small batches of flavored oil and
use within 24 hours. Keep the flavored
oil refrigerated when not in use.

Olive oil is one of those rare commodities—something that tastes wonderful and is actually good for you. Specialist food shops conduct olive oil tastings, like wine tastings. The oils are poured into small white dishes and accompanied by pieces of bread. The oils are tasted by dipping the bread into each oil and comparing the flavors. Oils at such **tastings** are offered labeled according to their country or region of origin, and also to their grade. The grades of olive oil are determined by the method of harvesting and pressing. The most desirable in terms of quality are hand-harvested, first pressed, and cold pressed. The finest oils from each region are produced in this way. The best oil is **extra-virgin**, with an acidity of less than one percent. It is best used as a flavoring agent—in dressings, or poured over finished dishes—so its inherent flavors are not altered by heat. Other grades include fine virgin olive oil, semi-fine olive oil, refined olive oil, and pure olive oil. You should choose a lower grade oil for cooking and for making mayonnaise. However, **aioli**, the rich garlic mayonnaise of Provence, should always be made from good, green, extra-virgin olive oil.

the flavors of
Olive Oil

Olives are grown all around the Mediterranean, as well as in South Africa, Australia, and California. The top oil producers are Spain, Italy, Greece, and France. Each country's oil has a distinctive **flavor**, whose nuances and color vary further by region—affected by climate, soil, and fruit variety. French oil is very **sweet** and, though produced in small quantities, is of fine quality. Greek oil is more **aromatic**, and oils from the Pelopponese and Crete are very highly regarded. Spain is the largest producer of oil and its oil tastes very **fruity**.

8 The flavors of olive oil

Fine Italian oils are produced in Liguria, Tuscany, Umbria, and Apulia. Tuscan oils have a **peppery** aftertaste, which varies from district to district. Californian olive oil is fruity and **mild**, and produced under varietal names such as Manzanilla, Sevillano, and Mission.

Australian oils are labeled, like their **wines**, with fruit varieties, climate, aspect of the groves, and pedigrees of the trees. Cyprus produces some superb oils, while oil produced in Israel, the Lebanon, Turkey, Syria, Tunisia, Portugal, Morocco, Algeria, and South Africa is mostly consumed in the home market.

The flavors of olive oil 9

Appetizers

Chilled tomato soup
with basil pesto

A soup for high summer when tomatoes are at their reddest and sweetest. Serve with garlic croutons—don't bother to cut them into neat dice—just tear the bread roughly into pieces, sauté with garlic in olive oil until lightly golden, then drain.

Place the tomatoes in a blender or food processor and puree to a mush. Pass through a stainless steel strainer into a bowl, pushing through as much juice and pulp as possible, giving a creamy consistency. Thin if necessary with the olive oil and tomato juice. Mix in all the remaining ingredients, place the bowl inside a larger bowl, and fill the space between the two bowls with ice. Pour cold water over the ice and chill in the refrigerator for about 2 hours to develop the flavors.

Serve in soup bowls, garnished with the basil pesto, strips of roasted yellow bell peppers, a drizzle of extra-virgin olive oil and an ice cube, together with a bowl of garlic croutons.

2 lb. very ripe plum tomatoes

3 tablespoons olive oil

⅔ cup tomato juice

3-inch piece of cucumber, finely chopped

1 yellow bell pepper, roasted, peeled, seeded, and chopped

1 small red onion, finely chopped

1 tablespoon balsamic vinegar

2 tablespoons torn fresh basil leaves

salt and freshly ground black pepper

to serve

basil pesto (see page 21)

roasted yellow bell peppers

extra-virgin olive oil

ice cubes

garlic croutons

Serves 4

Tuscan bean soup
with green pesto

A rustic soup, with a delicious green pesto swirled through just before serving. If fava beans are out of season, substitute green flageolet beans or more white cannellini beans. If you're short of time, you could also use canned cannellinis or flageolets.

To prepare the dried beans, soak overnight in cold water. Next day, drain, place in a large saucepan with cold water to cover, bring to a boil, and simmer gently until the beans are tender (about 30 minutes, depending on the age of the beans). Add a little salt about 5 minutes before the beans are done. To make the soup, heat the olive oil in a large saucepan, add the onion and cook over a medium heat until softened and transparent. Add the cabbage and cook for a few minutes. Add the stock, bring to a boil and simmer, covered, for about 20 minutes. Add the fava beans and zucchini, and simmer for a further 10 minutes. Stir in the oregano and season with salt and black pepper. Add the cannellini beans, with their cooking liquid, and simmer together for about 5 minutes. To serve, divide the soup between 4 large heated soup bowls, then place a tablespoon of pesto on top. Add a sprig of basil, if using, and sprinkle with the roughly crumbled Parmesan.

½ cup dried white cannellini beans

⅔ cup pure olive oil

1 onion, finely chopped

¼ cabbage, finely sliced

4 cups vegetable stock

1 cup fava beans (about 1 lb. in the pod)

2 zucchini, sliced and quartered

2 tablespoons snipped fresh oregano leaves

salt and freshly ground black pepper, to taste

to serve

4 tablespoons basil pesto (see page 21)

4 sprigs of basil (optional)

4 tablespoons crumbled fresh Parmesan cheese

Serves 4

a rustic Italian soup, with brilliant green basil pesto

Goat cheese
tomato and lentil salad

Bordeaux lentils have a nutty taste and stay whole and firm when cooked—their flavor contrasts well with the strong goat cheese. Puy lentils, darker in color, are more widely available, with a more robust flavour.

Place the lentils in a saucepan and cover with cold water. Add the garlic and half the parsley, bring to a boil and simmer for 10 minutes.

Add the onion, celery, and a pinch of salt, and simmer for 10 to 15 minutes more, adding a little hot water if required, until the lentils are just cooked but still firm. Discard the herbs and garlic, then toss the lentils in olive oil while still warm.

Crumble the goat cheese and mix into the lentils. Add the snipped chives, lemon juice, and the remaining parsley, chopped, then season to taste. To serve, place salad leaves on each plate, spoon over the lentil and cheese mixture, then garnish with the cherry tomatoes, chopped into chunks.

½ cup Bordeaux, Puy, or
other brown lentils

1–3 garlic cloves

1 bunch of fresh
flat-leaf parsley

1 red onion, finely diced

1 celery stalk,
finely diced

pinch of salt

1 cup virgin olive oil

8 oz. goat cheese

1 bunch of chives,
snipped

¼ cup lemon juice

salt and freshly
ground black pepper

to serve

salad leaves, such
as arugula, spinach,
watercress, or mâche
(lamb's lettuce)

8 oz. cherry tomatoes

Serves 4

a very **eas**

with truffle oil to give tha

Leek and new potato salad
with truffle oil

Truffle oil is the most important ingredient in this recipe, and is also delicious drizzled over pasta. Buy it in small quantities, such as the little bottles of white and black truffle oils pictured. Tiny baguette leeks look great in salads, but if you can't find baguettes, use larger leeks cut into chunks and leave the chunks intact when serving.

Cook the potatoes in boiling salted water for about 20 minutes, until tender.
Trim the leeks and slice into 1-inch chunks, if large. Wash thoroughly and cook in boiling salted water for 5 to 7 minutes, until just tender.
Place the parsley, oil, salt, and pepper in a bowl.
Drain the potatoes, slice thickly, and add to the dressing while still warm, so they absorb the flavors.
Drain the leeks thoroughly and add to the potatoes.
Stir well, taste, and adjust the seasoning.
Spoon onto serving plates, then drizzle with truffle oil, and serve.

1½ lb. new potatoes

1 lb. young leeks
(preferably baguettes)

2–4 tablespoons
roughly chopped fresh
flat-leaf parsley

3 tablespoons olive oil

sea salt and freshly
ground black pepper

4 tablespoons truffle oil,
to serve

Serves 4

ut **spectacular** salad—

xtra **dash of luxury**

Fried quail egg salad
with bacon and leaves

For the best look, flavor, and texture, fry the
eggs, but boiled eggs may also be used.
Nasturtium flowers make a colorful, peppery
addition to this salad in summer.

mixed leaf salad

**nasturtium flowers
(optional)**

4 tablespoons olive oil

8 quail eggs

**4 slices bacon,
cut crosswise into
½ inch strips**

**2 tablespoons
red wine vinegar**

**1 teaspoon
wholegrain mustard**

**salt and freshly
ground black pepper**

Serves 4

Arrange the leaves (and flowers if using) on
4 appetizer plates.

Heat 1 tablespoon of the oil in a small skillet
and lightly cook 4 of the eggs. Carefully place on the
leaves. Add another tablespoon of oil and repeat
with the remaining eggs.

Add the bacon pieces to the skillet, sauté until crisp,
then add to the salad.

Whisk in the remaining 2 tablespoons of olive oil,
together with the vinegar and mustard, then pour
over the eggs and salad leaves. Season with salt and
freshly ground black pepper

Serve with crusty bread.

Bruschetta
with tomatoes and olives

Bruschettas make perfect fast snacks—and also delicious appetizers for a summer lunch. Try other toppings too, such as the goat cheese on page 29, the artichokes on page 26, or the eggplant on page 58.

Split the focaccias in half and toast lightly. Spread with a dollop of green basil pesto, then add the tomatoes and olives. Add a slice of mozzarella to each, then drizzle with olive oil and place under the broiler for a few minutes. Serve, garnished with sprigs of basil leaves.

4 focaccias or sliced baguette loaf

green basil pesto (see opposite)

6 oz. vine-ripened cherry tomatoes

⅓ cup black olives, pitted and quartered

2–4 mozzarella cheeses, about 4 oz. each, sliced

Italian single estate olive oil

sprigs of basil, to garnish

Serves 4

Olive tapenade

If available, use olives and olive oil from the same region or country of origin.

Pit and chop the olives, then place in a blender or food processor with the capers, anchovies, and garlic, and blend thoroughly. Add the olive oil gradually until the mixture reaches a spreadable consistency. Slice and toast the baguette, spread with the tapenade, and serve with broiled tomato slices.

6 oz. black olives

2 tablespoons capers

4 anchovy fillets, rinsed and chopped

2 garlic cloves, crushed

⅔ cup olive oil

to serve

1 baguette loaf

broiled tomato slices

Serves 4

Green basil pesto

Ready-made pesto is widely available in shops and supermarkets, but it's never as good as homemade. Whenever you see some good fresh basil, or if you grow your own, try this recipe—it's a revelation!

Place all the ingredients in a blender or food processor and puree to a thick cream, adding a little more oil if it is too thick. Add salt to taste. Spoon into a jar, and float a thin layer of oil on the top. Store in the refrigerator until ready to use.

4 oz. fresh basil leaves

2 tablespoons pignoli nuts

¼ cup fresh Parmesan cheese, grated

⅔ cup olive oil

salt

Makes one ½-pint jar

Flavored oil

You can buy flavored oils, but it's simple to make your own. Make very small quantities, store them in the refrigerator and use them quickly. Sterilize both bottles and corks thoroughly before using. To avoid possible bacterial growth, it is best not to make garlic or truffle oils yourself.

If using herbs, wash them thoroughly before use. Pour the oil into a saucepan and gently heat together with the herbs or other flavorings. Pour the hot oil into sterilized bottles, then add the heated herbs or flavorings. Store in the refrigerator and use within 24 hours. *

virgin olive oil

flavorings

a choice of:
1 sprig of rosemary,
2 sprigs of thyme,
1 tablespoon black peppercorns,
2 sprigs of bay leaves,
2 pieces of lemon peel
or 1 tablespoon dried chiles

*Warning: please refer to FDA directive on page 4.

Left, front row, from left, Bruschettas with marinated artichokes (recipe page 26), with olive tapenade (page 20), and with green basil pesto (page 21). Right, with green basil pesto, and with tomatoes and olives (recipe page 20—see also back jacket).

Char-grilled asparagus
with a warm bread salad

The Italians use bread soaked in oil or
vinegar as a base for sauces and soups.
Char-grilling or roasting asparagus keeps
the spears crisp on the inside.
Caperberries are delicious if you can find
them—but ordinary capers also work well.

To make the salad, tear the bread into chunks and
mix with half the vinegar until softened. Squeeze dry
and mix with the egg, parsley, caperberries or
capers, salt, and pepper. Heat the oil and remaining
wine vinegar in a saucepan until warm, then stir in
the bread mixture. Keep the mixture warm while you
cook the asparagus.
Heat a grill pan on top of the stove and lightly brush
with olive oil. Add the asparagus spears and cook on
all sides until slightly shriveled and brown.
Alternatively, place in a roasting pan and cook in a
preheated oven at 400°F for 5 to 7 minutes until
browned. Place on a warmed serving plate, spoon
over the warm bread salad, sprinkle with the
balsamic vinegar, and serve immediately.

1 lb. asparagus

oil, for brushing

2 teaspoons balsamic
vinegar, to serve

**warm
bread salad**

3 oz. bread (sourdough
or country style)

⅔ cup white
wine vinegar

1 hard-boiled egg,
quartered

1 bunch of fresh
flat-leaf parsley,
roughly chopped

2 oz. caperberries,
(if available),
or capers packed
in salt or brine

⅔ cup virgin olive oil

salt and freshly
ground black pepper

Serves 4

Baby artichokes
marinated in olive oil and lemon

These wonderfully flavored artichokes
can be made in advance and stored in
sealed jars. Serve them in this salad,
as a bruschetta topping, or as an antipasto
with sliced salami.

Peel away the outer layer of the artichoke stems and
trim the leaves with a knife or scissors. (Young,
tender artichokes have no prickly chokes.)
Brush over all the cut surfaces with a little lemon
juice to prevent browning.
Place in a shallow saucepan with a tightly fitting lid.
Add the lemon juice, bay leaves, peppercorns,
parsley, olive oil, sea salt, and water to cover.
Bring to a boil and simmer for 10 to 15 minutes until
tender. Set aside to marinate for at least 24 hours.
When ready to serve, drain and reserve the liquid to
use again as a vinaigrette. Cut the artichokes in half,
bunch the spinach or mustard leaves on the toasted
bread, place the artichoke halves on top
and pour over some of the marinade.

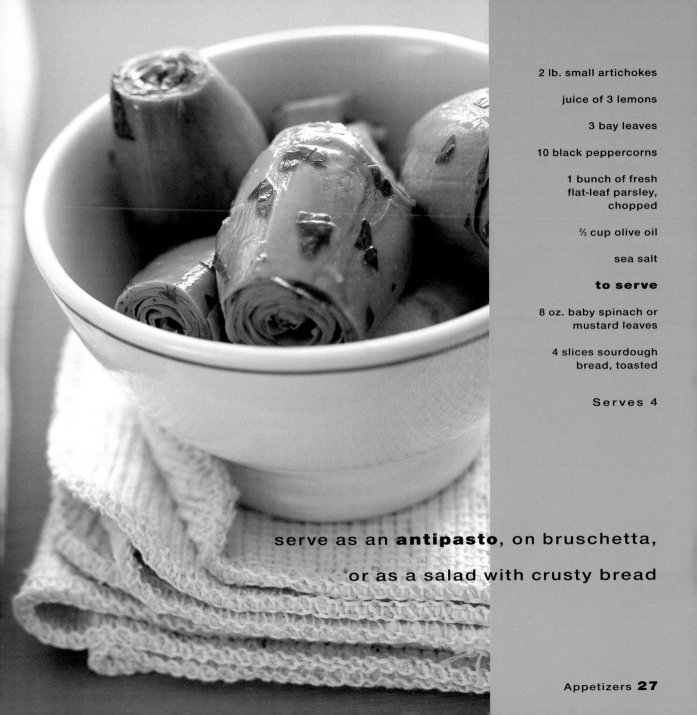

2 lb. small artichokes

juice of 3 lemons

3 bay leaves

10 black peppercorns

1 bunch of fresh
flat-leaf parsley,
chopped

⅔ cup olive oil

sea salt

to serve

8 oz. baby spinach or
mustard leaves

4 slices sourdough
bread, toasted

Serves 4

serve as an **antipasto**, on bruschetta,

or as a salad with crusty bread

Goat cheese
marinated in olive oil

Marinate your own goat cheese in olive oil—it can then be used broiled on toast, as a pasta sauce with snipped herbs, or served with a selection of crisp crackers.

Sterilize 2 small Mason jars by placing in a deep saucepan, pouring over boiling water and simmering for 15 minutes. When cool, place the goat cheese in the sterilized jars. Pour in the olive oil, add salt, then push in the olives, if using. Blanch the garlic cloves, herbs, spices, and lemon peel in boiling water for 1 minute, drain and add to the jar. Cover and leave to marinate in the refrigerator for 2 days before using, then use within 1 week.

1 lb. goat cheese, sliced and quartered

2½ cups virgin olive oil

½ cup pitted black olives (optional)

4 garlic cloves

a selection of:
1 small red chile,
2–3 sprigs of thyme or rosemary,
2 bay leaves,
and 1 tablespoon peppercorns

lemon peel

sea salt

**Makes two
½-pint jars**

Mushrooms and anchovies
with sautéed polenta wedges

great for lunch

Polenta can be bought ready made or prepared according to the recipe below. As an alternative, use slices of focaccia, drizzled with olive oil and baked in the oven until crispy. You can also make this recipe without the anchovies.

To make the polenta, place in a pan, add the boiling water, and prepare according to the package instructions. When done, spread into a deep pan to set (about 5 minutes), then cut into wedges and sauté in the olive oil until lightly browned. Alternatively, place the pieces in a baking pan, drizzle with a little olive oil, and cook in a preheated oven at 400°F for about 10 minutes. Rinse the porcini, soak in warm water for 20 minutes, then drain on paper towels, and roughly chop. Reserve the soaking liquid. Heat the oil in a skillet, and sauté the garlic for a few minutes until softened and transparent. Add the porcini and sauté for a few more minutes. Stir in the sherry, chopped anchovies, and porcini soaking liquid, bring to a boil, and simmer for about 5 minutes. Taste and adjust the seasoning. Remove the mushrooms, then bring the sauce to a boil and reduce to about 4 to 6 tablespoons. To serve, spoon the mixture onto the warm polenta, pour over the sauce, and sprinkle with oregano.

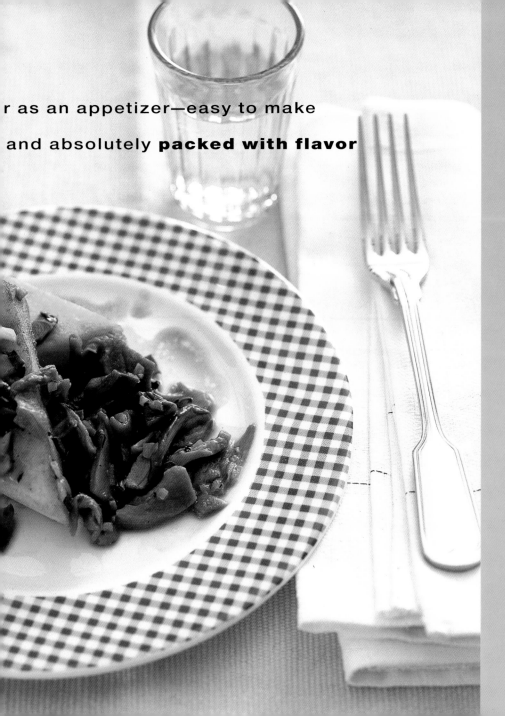

r as an appetizer—easy to make

and absolutely **packed with flavor**

8 polenta pieces or
homemade polenta
(see below)

about 4 oz. dried
porcini mushrooms

3 tablespoons
virgin olive oil

1 garlic clove, crushed

1 tablespoon dry sherry

2 oz. anchovies, canned
in olive oil, finely
chopped

sea salt and freshly
ground black pepper

chopped fresh oregano
leaves, to serve

fried polenta

1 cup pre-cooked
polenta

1 cup boiling water

olive oil, for sautéing

Serves 4

Entrees

Char-grilled tuna
with red onion salad and gremolata

Marinating is a way of giving the fish extra flavor. This recipe is perfect for cooking on a barbecue in the summer, but you can get a similar effect using a grill pan on top of the stove. This recipe is also good with other big fish such as swordfish.

To make the gremolata, mix the parsley with the olive oil, lemon juice and zest, salt, and freshly ground black pepper. Set aside.

To make the salad, place the sliced onions in a shallow dish, pour over the olive oil and balsamic vinegar, and sprinkle with the herbs and seasonings. Set aside to develop the flavors.

Brush the tuna steaks with the olive oil. Heat a grill pan on top of the stove and, when very hot, quickly sear the tuna on both sides. Add seasoning, then continue to cook for about 5 minutes. Divide the onion salad between 4 dinner plates with the tuna beside. Place a spoonful of gremolata on each steak, and serve with wedges of lemon.

4 tuna steaks

⅔ cup olive oil

salt and freshly ground black pepper

lemon wedges, to serve

gremolata

1 large bunch of fresh flat-leaf parsley, roughly chopped

½ cup olive oil

juice and finely sliced or grated zest of 1 lemon

salt and freshly ground black pepper

red onion salad

3 red onions, sliced

2 tablespoons extra-virgin olive oil

1 tablespoon balsamic vinegar

3 tablespoons finely chopped fresh flat-leaf parsley

salt and freshly ground black pepper

Serves 4

Roasted cod
with olive tapenade

The strong flavor of this tapenade works
very well with cod or other big fish that hold
their shape, such as haddock or flounder.

4 cod steaks,
about 4 oz. each

⅔ cup pitted
green olives

1 tablespoon capers
packed in salt, or in
brine (drained)

1 bunch of fresh
flat-leaf parsley,
chopped

8 anchovy fillets

3 tablespoons olive oil

sea salt and freshly
ground black pepper

steamed couscous,
to serve

Serves 4

Place the cod steaks in a roasting pan.
Mix the olives with the capers and parsley.
Spread this mixture on top of the cod steaks, place
2 anchovy fillets in a criss-cross on top of each
steak, pour over the olive oil, and season with salt
and freshly ground black pepper.
Cook in a preheated oven at 400°F for
about 15 to 20 minutes until the fish turns
white and opaque.
Serve with steamed couscous.

Seared beef fillet
with olives, tomatoes, and arugula

A spectacularly easy recipe, packed with
flavor and goodness. Cook the beef in a
grill pan on top of the stove, or very quickly
in a searingly hot oven.

Cut slits into the surface of the beef and insert the
garlic slices. Pour over the olive oil and sprinkle with
freshly ground black pepper. Tuck the bay leaves
underneath and set aside for a few minutes.
To prepare the arugula salad, place the leaves in a
bowl, add the tomatoes and olives, if using.
Drizzle with the extra-virgin olive oil.
Heat a grill pan on top of the stove until very hot.
Add the beef and bay leaves and cook the beef for
about 5 minutes on each side, depending on how
rare you like your beef.
If roasting in the oven, cook in a preheated oven
at 400°F for 10 minutes for rare beef, 15 minutes
for medium, and 20 minutes for well done.
Discard the bay leaves. Set the beef aside,
covered, for 5 minutes, to allow the meat to rest,
then cut into thick slices.
Place a pile of arugula leaves on each plate, top
with the beef slices, add the sliced tomatoes and
black olives, if using, then dress with a little more
extra-virgin olive oil and sprinkle with torn fresh
flat-leaf parsley and sea salt.

1 lb. beef fillet

4 garlic cloves,
sliced lengthwise

⅔ cup olive oil

4 bay leaves

sea salt and freshly
ground black pepper

to serve

1 large bunch
of fresh arugula

4 ripe plum
tomatoes, sliced

handful of black
olives, preferably
herbes de provence
(optional)

extra-virgin olive oil,
preferably from
Provence

1 tablespoon torn
fresh flat-leaf parsley

Serves 4–6

Provençal lamb daube

This is a quick and easy update of one of the great classics from the South of France. Use oil from Provence, if you can find it, for a truly authentic flavor.

Place the lamb in a shallow pan. Mix the marinade ingredients together and pour over the meat. Place in the refrigerator overnight or up to 2 days, depending on time available. Remove from the refrigerator about 1 hour before cooking to return to room temperature.

Remove the lamb from the marinade and pat dry with paper towels. Reserve the marinade.

Heat the olive oil in a large skillet and brown the meat on all sides.

Transfer to a casserole dish, sprinkle the flour over the oil and juices in the skillet, stir and cook for about 1 minute, then add the brandy.

Bring to a boil, remove from the heat, and pour over the meat in the casserole dish.

Pour in the marinade and stock, adding water to cover the meat if necessary, season well, and bring up to boiling point on top of the stove.

Transfer to a preheated oven and cook slowly at 325°F for 1 to 1½ hours.

Serve the daube with stewed flageolet beans and mashed potato made with olive oil.

6 large lamb shanks, left whole, or 1 leg of lamb, about 3 lb., cut into thick slices

3 tablespoons olive oil

2 tablespoons flour

½ cup brandy

1¼ cups vegetable or lamb stock

salt and freshly ground black pepper

marinade

4 garlic cloves

2 carrots

1¼ cups red wine

2 sprigs of thyme

2 sprigs of parsley

2 strips of orange peel

4 tablespoons olive oil

Serves 6

Marinated chicken
with avocado and spinach salad

It is important to use free-range chicken to give a good flavor in such a simple recipe— or in any other! Serve it either hot or cold, with a salad such as this one.

Place the chicken in a single layer in a roasting pan, mix the remaining ingredients together and pour over the chicken. Marinate in the refrigerator for at least 30 minutes, or longer if possible. Cook in a preheated oven at 400°F for 20 minutes, or until the juices run clear when the chicken is pierced with a skewer through the thickest part. To make the salad, blanch the beans in boiling water and refresh in iced water. Using a teaspoon, scoop out balls of avocado and place in a bowl with the beans, lettuce, and spinach leaves. Whisk the oil with the vinegar, salt, and freshly ground black pepper, pour over the salad and serve immediately with the chicken. A dish of boiled new potatoes would be a suitable accompaniment.

an easy dish for a midweek dinner party

6 pieces of free-range chicken, about 2 lb.

juice of 1 lemon

½ cup virgin olive oil

½ cup pitted green olives

1 tablespoon capers, packed in salt

1 bay leaf

4 sprigs of rosemary

avocado and spinach salad

4 oz. green beans

1 ripe avocado

1 escarole or other soft lettuce

4 oz. baby spinach leaves

⅔ cup extra-virgin olive oil

¼ cup balsamic vinegar

salt and freshly ground black pepper

Serves 4–6

Chicken breasts
with pancetta and rosemary oil

Rosemary oil is a wonderful flavoring ingredient. Use it for sautéing, roasting, or as a quick dressing for salads and steamed vegetables. Serve this dish with whole new potatoes, tossed in olive oil and crushed garlic, sprinkled with sea salt, then baked in the oven at the same time as the chicken.

4 skinless chicken breasts, free-range if available

4 slices Smithfield ham, or other high-quality ham

rosemary oil

½ cup extra-virgin olive oil

6 sprigs of rosemary

3 whole garlic cloves

pancetta filling

4 oz. pancetta, finely chopped

½ cup ricotta cheese

sea salt and freshly ground black pepper

To make the rosemary oil, place all the ingredients in a small saucepan, heat gently until tepid, then set aside for about 15 minutes until ready to use.
To make the filling, mix the pancetta with the ricotta, salt, and freshly ground black pepper.
Using a small, sharp knife, cut pockets down the sides of the chicken breasts and loosely pack with the filling. Wrap each breast in a slice of ham. Place the chicken in a roasting pan, pour over the rosemary oil, and bake in a preheated oven at 400°F for 20 minutes.
Serve immediately with oven-baked garlic potatoes and a green leafy salad.

Serves 4

Braised pheasant
with apples, fennel, and sage

A winter dish which also works well with chicken, guinea fowl, quail, or other birds. Omit the fennel if it is unavailable—the dish is equally good without it. You could also leave out the flour—the golden sauce will be thinner, but it will still taste terrific.

Heat the olive oil in a large skillet, add the pheasant pieces, and cook until golden brown on all sides. Transfer to an ovenproof casserole dish. Sauté the bacon until golden, then add to the casserole. Add the flour, if using, to the pan, stir well and cook for 1 minute until lightly browned. Stir in the apple juice, chicken stock, sage, and parsley, bring to a boil, then pour over the pheasant. Stir in the fennel, season, and bring to a boil on top of the stove. Transfer to a preheated oven and cook at 400°F for 45 minutes, or until just tender. Melt the butter in a small skillet and, when sizzling, add the apple slices and a pinch of sugar, and sauté until golden. Stir into the cooked pheasant, then serve with mashed potato made with crushed garlic, and a glass of warming red wine.

a winter dish with a **golden**

about ⅔ cup olive oil

2 pheasants,
cut into 8 pieces

6 slices of bacon, cut
into small pieces

1 tablespoon flour
(optional)

⅔ cup apple juice

1¼ cups chicken stock

2–3 large sprigs
of sage

1 large bunch of fresh
flat-leaf parsley

1 fennel bulb,
thickly sliced

2 tablespoons butter

2 apples, sliced

pinch of sugar

salt and freshly
ground black pepper

Serves 4–6

sauce that tastes terrific

Pasta

Pappardelle
with olive and parsley sauce

A good-quality dried pasta is best for this recipe, though fresh pappardelle from a very good Italian deli could also be used.

To make the sauce, heat 1 tablespoon of the olive oil in a small skillet, add the shallot, and sauté until softened and transparent. Add the remaining sauce ingredients, simmer for 10 minutes, then place in a blender or food processor and puree until smooth. Bring a large pan of salted water to a boil, add the pasta and cook according to the package instructions until just al dente. Drain and serve immediately in heated pasta dishes, with the sauce spooned over, then sprinkled with the chopped fresh oregano.

1 lb. good quality dried pappardelle

4 tablespoons chopped fresh oregano, to garnish

olive and parsley sauce

½ cup olive oil

1 small shallot, finely chopped

⅔ cup green olives, pitted

2 garlic cloves, crushed

4 sun-dried tomatoes, finely chopped

grated zest and juice of 1 lemon

1 bunch of fresh flat-leaf parsley

1 bunch of fresh oregano

salt and freshly ground black pepper

Serves 4

Baked green gnocchi
with creamy onion confit

A warming dish with sweetly flavored onions
in a cream sauce, baked with green spinach
gnocchi and Parmesan.

To make the onion confit, heat the olive oil in a large
pan, add the onions, then stir. Cover the pan with a
lid and cook until softened and golden.
Take care not to burn the onions, and add a little
more olive oil if necessary.
Add the garlic, cream, and eggs, and heat gently.
Meanwhile, cook the gnocchi in boiling salted water
for a few minutes until they rise to the surface.
Drain and stir into the sauce.
Spoon the mixture into a gratin pan, sprinkle with
the Parmesan, and cook in a preheated oven at
400°F for 15 to 20 minutes.
Sprinkle with pepper and serve immediately.

1 lb. fresh gnocchi,
spinach or plain

2 oz. freshly grated
Parmesan cheese

freshly ground
black pepper

onion confit

2 tablespoons olive oil

1 lb. onions,
finely sliced

1 garlic clove, crushed

⅔ cup light cream

2 eggs, lightly beaten

Serves 4

Pumpkin ravioli
with a pumpkin and dill sauce

Olive oil and chopped dill are used to flavor
both the sauce and the filling in this
gorgeous, golden ravioli recipe.

To make the filling, heat the oil in a skillet, and sauté
the pumpkin cubes very gently for about 20 minutes
until very soft. Add the dill, salt, and pepper.
To make the pasta, mix the egg and flour together,
adding enough water to form a stiff but smooth
dough. Knead for 10 minutes until very smooth.
Cover and rest the dough for 40 minutes. Using a
pasta machine, roll out the dough into a long strip
about 5 inches wide. Place spoonfuls of filling in
2 rows, about 1 inch from the top and bottom, and
the same distance apart, until half the pasta sheet is
covered. Moisten the edges and the spaces in
between with water. Fold over the other half of the
sheet, then flatten and seal the sections between the
fillings. Cut out the squares with a zigzag ravioli
wheel. Alternatively, use a ravioli cutter or a pasta
machine with the ravioli attachment.
Chill the filled ravioli while you make the sauce.
To make the sauce, heat the oil in a skillet, add the
pumpkin and shallot, and sauté until tender. Add the
dill, salt, and pepper. Cook the ravioli in a large pan
of boiling salted water for 4 to 5 minutes, until they
rise to the surface. To serve, pour into heated plates,
spoon over the sauce, and sprinkle with Parmesan.

1 egg

2 cups all-purpose flour

water, as needed

ravioli filling

2 tablespoons olive oil

1 lb. pumpkin or
butternut squash,
peeled and finely cubed

1 tablespoon
chopped fresh dill

salt and freshly
ground black pepper

**pumpkin and
dill sauce**

2 tablespoons
extra-virgin olive oil

8 oz. pumpkin or
butternut squash,
peeled and finely cubed

1 shallot or small onion,
finely chopped

2 tablespoons
chopped fresh dill

sea salt and freshly
ground black pepper

freshly grated
Parmesan cheese,
to serve

Serves 4

a **wonderful**, colorful sauce—served

over ravioli made with this delicious herbed pumpkin filling

Rice

Beet risotto

For fantastic flavor, roast the beetroot
yourself. This recipe is also delicious served
with a spoonful of sour cream.

If using fresh beets, roast them, unpeeled, in a preheated oven at 400°F for 30 minutes. Peel and dice.
To make the risotto, heat the oil in a saucepan, add
the onion and sauté until softened. Stir in the rice,
add a ladle of hot stock, stir, and allow the rice to
absorb the stock. Stir in the beets and more stock,
stir again, and simmer until absorbed. Repeat until
you have used all the stock. Taste, adjust the
seasoning and serve, topped with chopped sage
and shavings of fresh Parmesan.

8 oz. cooked beets,
peeled and diced,
or 1 lb. fresh beets,
unpeeled

2 tablespoons olive oil

1 small onion,
finely chopped

1½ cups risotto rice

1½ cups hot
vegetable stock

sea salt and freshly
ground black pepper

to serve

chopped fresh
sage leaves

shavings of fresh
Parmesan cheese

Serves 4

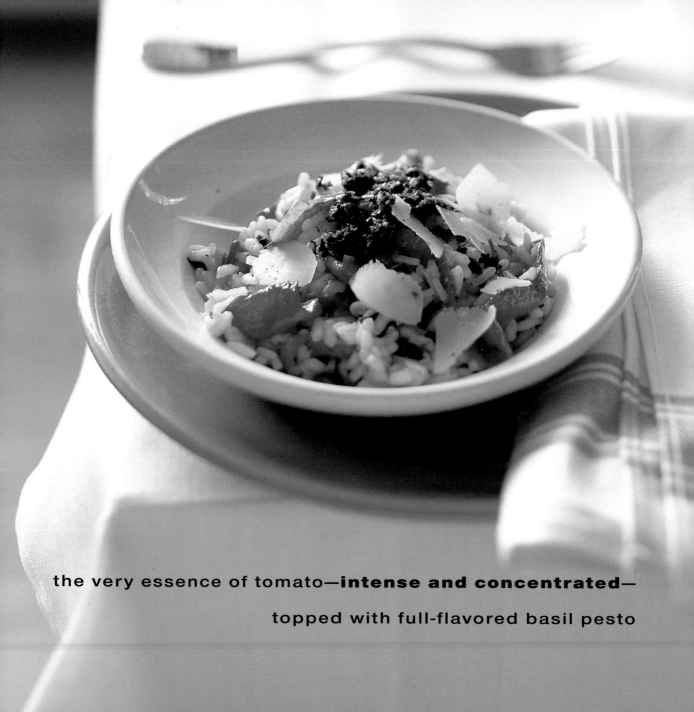

the very essence of tomato—**intense and concentrated**—

topped with **full-flavored basil pesto**

Baked tomato risotto
with green pesto

Although there are lots of wonderful flavors
for risotto, tomato and pesto are especially
good. Whip up your own pesto when basil is
plentiful, using the recipe on page 21.
Otherwise, use good-quality ready-made
pesto. The dense texture of plum tomatoes is
best suited to this recipe—and it's simple to
bake them in advance, very slowly, in a low
oven, so their liquid evaporates leaving an
intense, concentrated flavor.

1 lb. ripe plum
tomatoes, sliced

1 garlic clove, crushed

2 tablespoons olive oil

1 onion, finely chopped

1½ cups risotto rice

1½ cups hot
vegetable stock

salt and freshly
ground black pepper

to serve

4 tablespoons basil
pesto (see page 21)

shavings of fresh
Parmesan cheese

Serves 4

Place the tomatoes in a roasting pan with the garlic,
sprinkle with salt and freshly ground black pepper,
then bake in a preheated oven at 325°F for 1 hour.
Heat the olive oil in a saucepan, add the onion and
lightly sauté until golden. Stir in the rice, add a ladle
of hot stock, then stir and allow the rice to absorb
the stock. Add more stock, stir, and simmer until
absorbed. Repeat until you have used all the stock.
Taste and adjust the seasoning.
Serve, topped with the baked tomatoes, a spoonful
of basil pesto and some shavings of fresh Parmesan.

Vegetables

Mediterranean ragoût
of marinated vegetables

Choose your own selection of vegetables, according to what's in season. Baby artichokes and fava beans are especially delicious additions.

Blanch the zucchini in a pan of boiling, salted water. Remove with a slotted spoon and set aside. Add the onions and boil until soft and tender, then remove with a slotted spoon and set aside.
Blanch the mushrooms in boiling water, then drain. Place the marinade ingredients in a large pan, bring to a boil, add the blanched vegetables and simmer for 10 minutes. Cool, transfer to a bowl, cover, and refrigerate for a few days for the flavors to develop. Drizzle with extra-virgin olive oil and serve, accompanied by crusty bread.

1 lb. small zucchini

8 oz. pearl onions

2 cups button mushrooms

extra-virgin olive oil, to serve

marinade

2½ cups water

juice of 3 lemons

1¼ cups olive oil

2 sprigs of thyme

1 bunch of fresh flat-leaf parsley

4 bay leaves

1 small celery stalk, with leaves

10 black peppercorns

10 green olives

Serves 4

Pickled eggplants
for bruschetta or pasta sauce

A useful bruschetta topping or pasta sauce,
sprinkled with chopped sun-dried tomatoes.

Place the shallot, bay leaves, salt, vinegar, and water
in a pan and bring to a boil. Add the eggplants. Put a
plate on top to keep them submerged, simmer for
10 minutes until soft, then drain. Pour boiling water
over two 1-quart Mason jars and place in the oven at
300°F for 5 minutes. Remove, drain, and pack in the
eggplants, garlic, chiles, and rosemary. Cover with
oil, leaving ½ inch headroom. Seal, set in a roasting
pan filled with water, and place in the oven at 300°F
for 15 minutes, or in a boiling water-bath canner
according to USDA guidelines. Cool, chill for 2 days
before eating, and use within 1 month.*

1 shallot

2 fresh bay leaves

2 teaspoons salt

1¼ cups vinegar

2½ cups water

2 lb. baby eggplants
(or large ones cut
into chunks)

8 whole garlic cloves

2 fresh red chiles

4 sprigs of rosemary

olive oil, to cover

**Makes two
1-quart jars**

Roasted bell peppers
marinated in lemon-flavored oil

A great bell pepper recipe—ideal on its own,
on bruschetta or with pasta.

Broil the bell peppers on all sides until the skins are
blistered and blackened. Wrap in plastic wrap for
3 to 5 minutes. Slip off the skins, pull out the stems
and seeds, cut in half, and place in 4 glass jam jars.
Add the lemon peel, then pour over olive oil to cover.
Chill 2 to 5 days before using and use within 1 week.

8 red or yellow
bell peppers

peeled rind of 1 lemon

about 1¼ cups
virgin olive oil

**Makes four
½-pint jars**

*Please refer to the
directive on page 4.

Spiced garbanzos
with eggplant and tomatoes

Olive oil, used as a braising medium, gives
the garbanzos loads of flavor. If you're short
of time, canned garbanzos make this Middle-
Eastern-influenced dish very quick and easy
to prepare. Serve with saffron rice and
dishes such as pan-fried lamb fillet.

Heat the olive oil in a large skillet, add the chopped
onion and the garlic, and sauté gently until
softened and transparent.
Stir in the remaining ingredients, bring to a boil and
simmer gently until the tomato sauce has slightly
thickened. Serve immediately.

6 tablespoons
extra-virgin olive oil

1 onion, finely chopped

1 garlic clove, crushed

½ teaspoon turmeric

small pinch of
ground coriander

small pinch of
ground cumin

1 eggplant,
unpeeled, diced

1 lb. plum tomatoes,
blanched, peeled,
seeded and
finely chopped

2 tablespoons
red wine or stock

1½ cups garbanzo
beans, freshly cooked
or canned

sea salt and freshly
ground black pepper

Serves 4

a great main dish for vegetarians—and

also **delicious** as an appetizer

Orange and lemon
polenta cake

Olive oil can be used to replace butter in cakes, and the type of oil you use will create the flavor. If you don't want it too strong, start with a light oil and move on up to a strong Spanish or Greek oil. Polenta gives the cake a lovely, nutty texture and a pretty green-gold color. This cake is wonderful served plain with coffee, or as a dessert with crème fraîche and a glass of sweet white wine.

Place the orange and lemon in a saucepan with cold water to cover. Bring to a boil and simmer for about 30 minutes until very soft. Drain and reserve the fruit, then cool. Place in a blender or food processor and puree until smooth.

Beat the eggs with an electric mixer until pale and fluffy (about 5 to 10 minutes), then whisk in the sugar. Mix the flour with the baking soda and salt, stir into the egg mixture, then mix in the olive oil. Fold in the polenta and the pureed fruit.

Pour the batter into an 8-inch springform pan. Cook in a preheated oven at 350°F for about 50 minutes, until a skewer, inserted into the center of the cake, comes out clean. Remove from the oven, place on a wire rack, sprinkle with confectioners sugar and, when cool, remove from the pan. Serve plain, or with whipped cream or crème fraîche.

. . . and som

...hing sweet

1 orange

1 lemon

4 eggs

½ cup sugar

1¾ cups
all-purpose flour

3 teaspoons
baking soda

pinch of salt

1¼ cups pint olive oil

⅔ cup polenta or
yellow cornmeal

to serve

confectioners sugar

4 tablespoons
crème fraîche,
(optional)

Serves 8–10

Index